THE STARRY SKY

ROSE WYLER
PICTURES BY STEVEN JAMES PETRUCCIO

JULIAN MESSNER

TO THE READER: Remember never to look directly at the sun.

Text copyright © 1989 by Rose Wyler. Illustrations copyright © 1989 by Steven James Petruccio. All rights reserved including the right of reproduction in whole or in part in any form. Published by Julian Messner, a division of Silver Burdett Press, Inc., Simon & Schuster, Inc., Prentice Hall Bldg., Englewood Cliffs, NJ 07632.

JULIAN MESSNER and colophon are trademarks of Simon & Schuster, Inc. Design by Malle N. Whitaker. Manufactured in the United States of America.

(lib.) 10 9 8 7 6 5 4 3 2 1

(pbk.) 10 9 8 7 6 5 4 3 2

Library of Congress Cataloging-in-Publication Data
Wyler, Rose.
The Starry Sky/Rose Wyler; pictures by Steven James Petruccio. p. cm.—(An Outdoor science book)
Summary: Simple text introduces characteristics and properties of the sky, sun, stars, planets, and moon.
1. Astronomy—Juvenile literature. 2. Sky—Juvenile literature. [1. Astronomy. 2. Sky.] I. Petruccio, Steven James, ill. II. Title. III. Series: Wyler, Rose. Outdoor science book.
QB46.W9484 1989   520—dc19
88-31192
CIP   AC
ISBN 0-671-66345-3 (lib. bdg.)
ISBN 0-671-66349-6 (pbk.)

The author and publisher thank Clint Hatchett, of the American Museum of Natural History, for his helpful suggestions.

Why Night Is Dark

All day long, the sun lights up the sky.
Clouds may hide the sun,
but its light goes through them.
When the sun sets, day ends.
The sky loses its blue color
and dark night begins.

Why is the sky blue in the daytime?
The sun's light passes through the air.
Some of the blue light from the sun
is scattered by the air so the sky looks blue.
To see the blue color, you must look
through a lot of air.
When you look up on a clear day,
you are looking through miles and miles of air.
So you see the wonderful blue of the sky.

The blue of the sky starts to disappear
as the sun sets.
That's because color shows only in the light.
In the dark, the sky looks almost black,
and black is not a real color.
When the sun is setting, watch for colors
in the sky around it.
While the rest of the sky is getting dark,
colors around the sun are still bright and clear.

In the city
the sun seems
to set behind houses.

In the country
the sun seems
to set behind hills.

Near the sea
it looks as if the sun
sinks into the water.

Yet the sun stays in the sky.
It is always there,
always shining.
After the sun sets,
it shines on a part of
the Earth you cannot see.

From the ground you cannot tell that
the Earth is shaped like a ball.
You cannot see around it to the other side,
and that is where the sun is shining.
It is day there while your side has night.
If you were an astronaut, you could see
how the sun lights up the big, round Earth.
Sunlight cannot go through the Earth.
So only the side facing the sun has day.

Day begins on your side of the Earth when the sun rises in the east.
By noon, the sun is high in the sky.
Then the sun moves west until it sets.

Day after day you see the sun rise and set.
Yet the sun does not really move.
It seems to cross the sky
because the Earth is turning.
While the Earth turns you one way,
you see the sun move the other way.

◆◆◆◆

Your eyes fool you, just as they do when you look at a wall and turn your head. Turn one way and the wall goes the other way, or so it seems.

The Earth turns all day long and all night long. It makes one full turn every twenty-four hours, and in that time each side has a day and night.

◆ ◆ ◆ ◆

To see how this happens, shine a light on a ball or globe in a dark room. Pretend the light is the sun shining in space and the ball is the Earth.

To mark your side, stick a paper doll on it. Now turn the make-believe Earth around once.

While your side faces the sun, you have day, and the other side has night.
Then, the other side has day and you have night.
One side, then the other, faces the sun as the Earth turns round and round.

You never feel that the Earth is moving,
for it moves without stops, starts or jolts.
Things around you are all moving at the same rate.
So you cannot tell they are turning too.
But you can find signs that the Earth is moving.
The signs are in the sky, every day and night.

When the Stars Come Out

The stars, like the sun,
are always in the sky
and they are always shining.
In the daytime the sky is so bright
that the stars do not show.
But when the sky darkens,
there they are.
What are the stars, you wonder,
and why do they twinkle?

Stars are huge balls of hot, hot gas.
They are like the sun but they look small
because they are much, much farther away.
They are trillions and trillions of miles away,
shining in black space, high above the air.
Space is empty and does not move.
Stars do not twinkle there, but
twinkling begins when starlight hits the air.
The air moves and tosses the light around.

On a clear night, how many stars can you see?
More than you can count—maybe 3,000.
Yet you never see the whole sky.
In the northern part of the world, you see
different stars than in the southern part.

In the North, on clear nights, you can see a group of stars called the Great Bear. How it got that name no one knows for sure. Part of this group looks like a big dipper and many people call it that—the Big Dipper. Nearby is the Little Bear, and Little Dipper. The last star in it is the North Star. When you see it, you are facing north.

North Star

The Dippers seem to turn slowly round the North Star during the night.

◆◆◆◆

Ask a grownup to help you draw the Dippers with chalk on the inside of an umbrella. Draw the North Star at the center.

Now hold the umbrella over you and imagine it is the sky. Turn it and the Dippers turn around, just as they do in the sky.

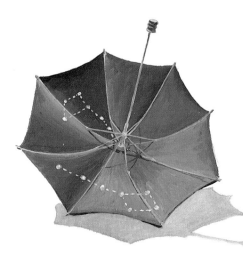

◆◆◆◆
Do other stars go around the sky too? Some night, pick a bright star that seems to touch the top of a pole or tree. Mark the spot where you stand and go back there an hour later to check your star. You will find it is not in the same place.

Try other stars and you will find they shift too. Like the sun, the stars seem to move because the Earth is turning.

The Earth also travels around the sun.
Each trip takes a whole year.
Along the way different stars come into sight.
From November to May the brightest star
is in a group called the Big Dog.
Nearby is Orion the Hunter, fighting the Bull.
A red star marks the Bull's angry eye.

The brightest star

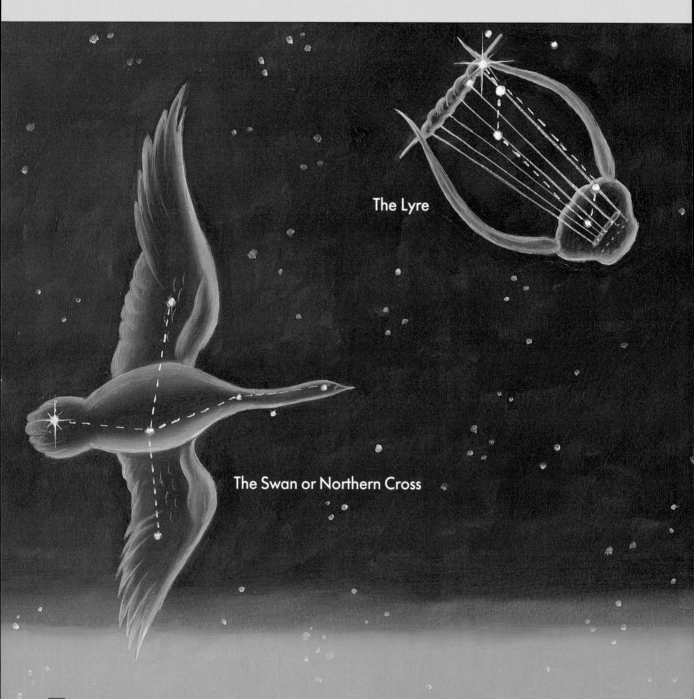

From May to November you can see the Lyre, the Swan or Northern Cross, and the Eagle. These groups are big and bright and easy to find. They lie along a shining band—the Milky Way.

The Eagle

With field glasses, you see that the Milky Way is made of stars, millions of them.
With a big telescope, billions more can be seen.
No wonder scientists say that there are more stars in the sky than grains of sand on all the seashores of the world!

Sometimes you see the bright Evening Star.
It shines in the west, near the setting sun.
When it is near the rising sun, it is called
the Morning Star. Yet it is not a star.
It is a planet, usually Venus.
Our Earth is a planet too.
Planets go around the sun and get light from it.
Besides Venus and the Earth,
seven other planets are known.

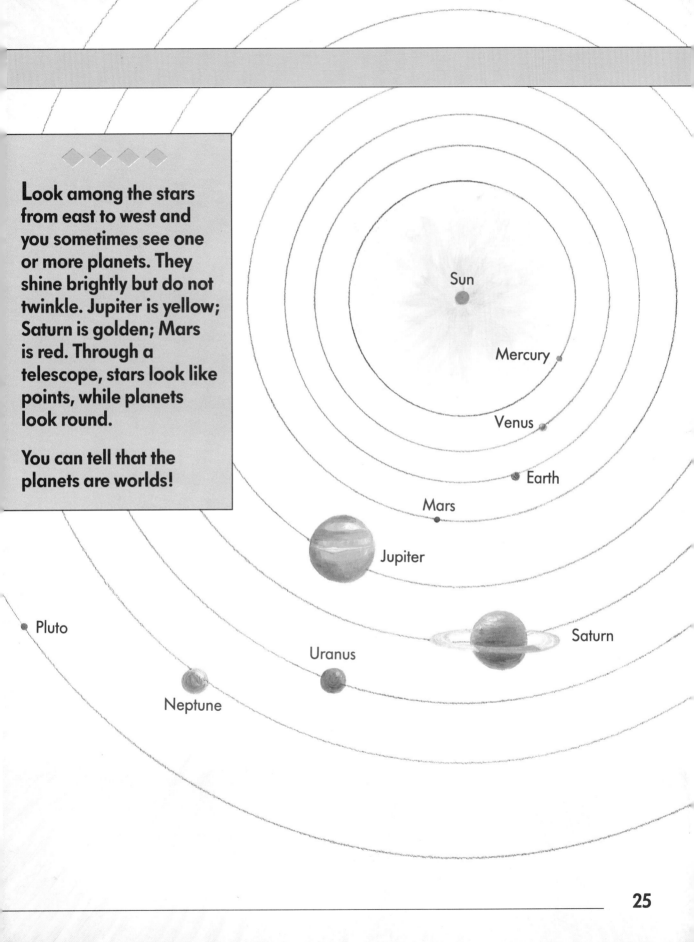

Look among the stars from east to west and you sometimes see one or more planets. They shine brightly but do not twinkle. Jupiter is yellow; Saturn is golden; Mars is red. Through a telescope, stars look like points, while planets look round.

You can tell that the planets are worlds!

25

Watching the Moon

Fewer stars show when the moon shines at night.
The moon is so bright that you can see it even
in the daytime while it seems to move across the sky.

How big the moon looks
when it rises!
Climbing up the sky,
the moon seems to get
smaller and smaller.
Then on the way down,
the moon seems to get
bigger and bigger,
until finally it sets.

But does the moon's size really change? Here's a way to find out. Hold a small button out in front of you and move it until it just covers the moon. Try this when the moon is low in the sky and when it is up high. No matter how big or small the moon looks, the button covers it. So you see the moon's size stays the same. It only seems to change.

The full moon looks like a face, doesn't it?
Dark areas look like eyes, a nose and mouth.

A telescope shows those areas are plains.
High mountains stand around the edges.
Here and there are hollows called craters.

Astronauts who have been to the moon find
it is a rocky place without air or water.
The sun and stars shine together in a black sky.

Sunlight does not go through the moon.
So it has a dark side and a sunny side.
You see all of the sunny side
as you look at the full moon.

When you see a half-moon or a crescent, you see only part of the moon's sunlit side. Why does that happen?

◆ ◆ ◆ ◆

Try this some sunny day when you see the moon. Let's say you see a crescent. Face it and hold up a white ball so that one side of it is lighted by the sun. You see only a part of that side and the ball looks like a crescent.

The moon moves around the Earth, changing from crescent to half-moon, to full moon, to half-moon, then to crescent again. The entire trip takes about a month.

◆◆◆◆
Now move the ball around so that you see more of its sunlit side. Watch the crescent become a half-moon.

Keep moving the ball in a circle and you see it change the way the moon changes in the sky.

How will the moon look tonight?
What stars can you see?
Will you see Venus or some other planet?
Just think—the Earth is a planet too,
moving around the sun
that is shining in black space
among billions of stars.
Tonight, when darkness comes,
what will you see
in the wide, wonderful starry sky?